# Creation, Evolution, and the Age of the Earth

### by Wayne Jackson

**Creation, Evolution, and the Age of the Earth, 2nd Edition**
© 2003 Courier Publications

All rights reserved. No part of this book may be reproduced in any form without written permission of the publisher except in the case of brief quotations embodied in critical articles or reviews.

ISBN 0-9678044-8-5

Additional copies may be ordered from:

**Courier Publications**
7809 N. Pershing Avenue
Stockton, CA 95207

http://www.courierpublications.com

# Table of Contents

Preface ........................................... 1
Introduction ...................................... 3
Chapter 1 ......................................... 5
*Father Time: Hero of the Plot*
Chapter 2 ........................................ 13
*Earth-Dating Methods: Concessions from Science*
Chapter 3 ........................................ 17
*Earth-Dating Methods: Assumptions Required*
Chapter 4 ........................................ 23
*Evolutionary Clocks: In Need of Repair*
Chapter 5 ........................................ 29
*The Bible and Inspired Chronology*
Chapter 6 ........................................ 35
*Biblical Genealogies and Human History*
Chapter 7 ........................................ 45
*Man: "Johnny-come-lately" or "From the Beginning"?*
Chapter 8 ........................................ 59
*Theories That Torture the Scriptures*
Chapter 9 ........................................ 71
*Physical Evidences for a Young Earth and Humanity*
Chapter 10 ....................................... 87
*Aguments for an Ancient Earth Answered*
Chapter 11 ...................................... 103
*A Concluding Word: Where is the Evidence?*
Endnotes ........................................ 109
Appendix ........................................ 113
*Why People Believe in Evolution*

The latest edition of this volume
is affectionately dedicated to:

## *David & Phyllis Sain*

Whose special friendship in recent years
has been a kindly gift of Providence
and a treasured blessing in the author's life.

# Preface

The theory of organic evolution is an insidious view that seeks to provide a naturalistic explanation for human existence, as a replacement for the biblical narrative. The motive behind the movement is transparent. It is rebellious man's attempt to cut the "ties that bind," thus launching out on his own, in terms of religious and moral conduct.

A careful examination of all the data, however, will reveal – to the objective student – that Darwin's theory is supported neither by Scripture, science, nor reason. Why, then, do so many subscribe to the evolutionary tale? Multitudes are intimidated by skeptical propaganda, so well orchestrated by various media outlets.

A key factor in the creation-evolution controversy is that of time. The Darwinists allege that "given enough time anything is possible" – even the seemingly absurd notion that the entire world of living creatures has evolved by natural processes from an initial sub-microscopic spark of life. What most fail to recognize, though, is that "time" has no intrinsic power. Without an appropriate mechanism for evolutionary change, time is meaningless. This book does not attempt to address the various arguments that are employed in the defense of evolutionary theory. It does look at the evidence for the "time" factor.

The idea that Earth's history has required vast eras of time is believed by a majority of people in our society. Even numerous folks who profess an identification with the Christian faith have accepted this viewpoint. Little do they realize the origin of this ideology. In reality, "time" does nothing for evolution. Without "time," though, by everyone's admission, the theory hasn't a ghost of a chance. Hopefully, this little volume will help thrust the importance of understanding "time" and it's relationship to the question of origins into focus.

# Introduction

As mentioned in the "Preface," so many unsuspecting people are intimidated by the evolutionary assertions regarding "time." This book, therefore, will enable the student to get a grasp on some of the fundamental issues relative to the importance of "time" in the evolutionary scheme.

Purposely, this volume has been designed to be brief and simple. Students are often discouraged by voluminous and complicated productions. From this abbreviated treatise, the average student will be able to carry away neatly-packaged, usable information, which not only will strengthen his own faith, but also will enable him to assist others who are struggling with the "time" assertions of the evolutionary community.

This study will look at "time" from two basic vantage points. It will examine the issue of chronology from the *biblical* perspective. Regardless of the superficial claims of some, chronology is an important Bible theme, and it must not be ignored. We will demonstrate, therefore, that the scriptural concept of time is absolutely adverse to the evolutionary time scale.

Further, though, we will examine the methods behind the Darwinian claims for a very ancient Earth (billions of years), and highlight the highly questionable assumptions upon which these techniques are based.

Also, we will look at the *science* data. Do the facts of science lend their support to the concept of a very ancient earth? Or do they, in fact, point to a rather recently-created planet – to be measured in terms of thousands of years, rather than billions?

This effort will also give consideration to several of the misguided attempts to harmonize the Scriptures with the evolutionary time scale, on the part of those who ostensibly profess confidence in the Bible. Such compromises, however, while popular with those who have not studied the issues carefully, are extremely disrespectful of the Sacred Text.

# 1 Father Time: The Hero of the Plot

Evolutionists are disagreed about many of the components so essential to their scheme. They argue about the *mechanism* that allegedly produced the world of living creatures. They dispute concerning *where* the phenomenon supposedly commenced – here on earth, or in outer space. They debate regarding the supposed *rate of development*; was it slow or rapid? One might almost state that there is virtually a civil war in the evolutionary camp.[1]

Be that as it may, however, there is certainly one matter upon which all of Darwin's disciples are agreed. It is this: *Time* is the essential element of the evolutionary scenario. Unless there is sufficient time, evolution hasn't a ghost of a chance of working.

One of the world's foremost scientists, hardcore evolutionist Dr. Robert Jastrow, expresses it like this: "The key to Darwin's explanation is time, and the passage of many centuries."[2]

Professor George Wald, for many years associated with Harvard University, in an article on the origin of life, paid tribute to "time" in the following fashion:

> "The important point is that since the origin of life belongs in the category of 'at-least-once' phenomena, time is on its side. However improbable we regard this event, or any of the steps it involves, given enough time, it will almost certainly happen at least once. And for life as we know it, with its capability for growth and reproduction, once may be enough. Time is the hero of the plot. The time with which we have to deal here is of the order of 2 billion years. What we regard as impossible on the basis of human experience is meaningless here. Given so much time, the impossible becomes possible, the possible becomes probable, and the probable becomes virtually certain. One has only to wait, time itself performs the miracles."[3]

During the 19th century scientists estimated the age of the earth at between 3 million and 1.6 billion years, based upon their speculations regarding sedimentation rates of the earth's strata. The increasing need for time, demanded by the theory of evolution, compelled them to assume longer and longer periods for earth's history. Accordingly, it is interesting to note that *over the past century, the estimated age of the earth has doubled approximately every twenty years!*

But even this has not been enough. For example,

the late Dr. George G. Simpson, an American champion of evolutionary dogma, in discussing "mutations" (i.e., genetic alterations) as a possible mechanism for explaining the evolutionary process, admitted that if there was an effective breeding population, of, say, 100 million individuals, and they could produce a new generation daily, the likelihood of obtaining good evolutionary results from mutations could be expected only about once in 274 billion years.[4] This, needless to say, is slightly beyond the currently estimated 5 billion-year-old earth!

What many people fail to realize, however, is the fact that time is merely *quantitative*, it is not *qualitative*. It has no creative genius, nor raw power whatever. Zero times zero for a billion years will never equal anything but *zero!*

The fact is, time contributes to degeneracy. If one stands on the edge of a cliff and flaps his arms for weeks, months, and years, in the hope that, given enough time, he will be able to fly, he will be sorely disappointed. Not only will he never fly, eventually, he will die of exhaustion! Time works *against* his ambition, not *for* it.

Or, consider this analogy. Suppose that one dropped one thousand 8 x 10 inch sheets of paper from an airplane at an elevation of 5,000 feet. What are the mathematical odds that, upon landing upon the earth, those sheets of paper would spell out the phrase, "In the beginning, God"? The obvious reply has to be: the odds against such happening would be in the billions. The papers would be scattered all over the countryside.

What if the experiment were repeated with the airplane flying at 10,000 feet (thus giving the paper more *time en route* to the earth)? Are the odds against the papers forming that phrase going to be greater or less? Obviously, the odds against such would be far greater, for the papers would be scattered over an even wider region. Time would not facilitate the formation of the phrase; rather, it would work against it.

According to the evolutionary scheme of things, the universe (via the "Big Bang") exploded into existence some 20 billion years ago. Subsequently, our earth was formed approximately 4.5 to 5 billion years ago. Then, as Wald observed, biological life allegedly was spontaneously generated about 2 billion years back, while man, a sort of Johnny-come-lately, arrived upon the scene some 3.6+ million years ago. Some are now even suggesting that human-kind has been around for possibly 6 million years.

There are three basic views that one may assume with reference to such matters.

1) One may assert, as evolutionists do, that the above figures are fairly well established and, therefore, the matter is beyond any reasonable controversy.

2) Some religionists, superficially claiming to believe the testimony of the Bible, and yet feeling the intimidating pressures of evolutionary propaganda, vacillate relative to this issue. They suggest that the time factor is really *irrelevant.* Such folks argue that the Bible is silent on this matter, thus, the universe may have been created billions of years ago, or a few thousand years ago; it simply does not matter.

This view can only be valid if the major assumption is true, i.e., that the Bible is silent on this issue. If the Scriptures *do* contain chronological data that speak to this theme, we must accept their testimony, regardless of the unproved assumptions of some modern scientists.

3) There are many people, including this writer, who believe that there is strong evidence – both biblical, historical, and scientific – that argues for a *recent creation* (measured in terms of *thousands* of years, not millions), and that to surrender this ground is to compromise the clear testimony of the Word of God. Moreover, it deprives us of the one clear point of attack against the evolutionary theory which every Darwinian concedes is vital to the system – that of *time.*

In the following pages we propose to show:

1) The dating techniques employed to yield vast eons of time in support of evolutionary chronology are laced with non-provable as-

sumptions. These assumptions are not only unfounded, but actually, there are valid reasons for doubting their reliability.

2) There are strong biblical arguments in support of the position that the creation occurred in the relatively recent past – certainly not billions of years ago.

3) There are reasonable historical and scientific evidences that buttress the biblical implications, and which suggest a recent creation.

# 2 Earth-Dating Methods: Concessions from Science

Is there actually "scientific evidence" that the earth is *billions of years old?* The average person will unhesitatingly answer, "Yes," to that query. Yet, if his very life depended upon it, he could not cite one scrap of proof supporting such. All that most folks know is this: "The scientists say so." True, but scientists say a lot of things that are not factual. One must always keep in mind that many scientists are evolutionists, and they have a vested interest in the "time" business.

Though the average layman assumes that geologists have firmly established the antiquity of the earth, the truth is, they have not; and the more candid ones will admit this. Dr. Stephen Moorebath, of the University of Oxford, an evolutionist, wrote:

> "No terrestrial rocks closely approaching an age of 4.6 billion years have yet been discovered. The evidence for the age of the earth is circumstantial, being based upon ... indirect reasoning."[5]

Dr. John Eddy, of the High Altitude Observatory in Boulder, Colorado, in an article which appeared in the prestigious journal **Geotimes**, declared:

> "There is no evidence based solely on solar observations that the Sun is 4.5 to 5 billion years old ... I suspect that the Sun is 4.5 billion years old. However, given some new and unexpected results to the contrary, and some time for frantic recalculation and theoretical readjustment, I suspect that we could live with Bishop Ussher's value for the age of the Earth and Sun. I don't think we have much in the way of observational evidence in astronomy to conflict with that."[6]

Ussher calculated the creation date at 4004 B.C., and though conservative Bible scholars no longer strictly hold to Ussher's dating system, it is quite significant that an evolutionary astronomer can concede that there is really no scientific evidence that is inconsistent with a 6,000 year old earth.

Dr. Robert Kofahl has noted that "it is not possible to 'prove' that the earth is billions of years old."[7] Or in the words of Dr. Donald Chittick, "the idea that the earth is very, very old is not in any way suggested by any studies in science. It arises as a result of rejecting Special Creation."[8] He is absolutely correct.

The next time someone tells you that the earth is

billions of years old, ask him: "How do you know?"

If he responds, "Well, there are dating methods that prove the earth is billions of years old," ask: "Could you list some of them for me?"

If he has not walked off by then, and if, by chance, he is able to mention one of the techniques (e.g., uranium-lead), ask him: "Are these methods absolute, or are certain assumptions built into them?"

If he is an honest individual, he will concede at this point that these dating procedures are not absolute by any means, and therefore, that they do not establish a provable date for the earth.

We will examine some of these dating assumptions in the chapter to follow.

# 3 Earth-dating Methods: Assumptions Required

Many people are aware that various techniques are employed to arrive at the presumed age of rock samples from the various strata of the earth. What most folks do not appreciate is the fact that these earth clock methods are all grounded in certain assumptions that absolutely must be granted in accepting the chronological data yielded by them. Let me attempt to explain, in a very simple fashion, the type of method generally employed in the dating game.

Radiometric methods for dating the earth's rocks are based upon the decay sequences of certain elements. For example, uranium-238 (called a parent element) will, through a series of decomposition processes, ultimately produce lead-206 (called a daughter element). Since scientists believe they know the present rate of decay, if a sample rock is found to contain both uranium-238 and lead-206, the ratio of the two is used to determine the age of the sample.

However, scientists admit that in order for this technique to be valid, certain *assumptions* must be granted, and, if it should turn out that these are *false* assumptions, then the entire procedure would be worthless. In our illustration of the uranium-lead method, the following assumptions are made.

a) Assumption One – There was *no* lead-206 in the rock at the time of its formation.

b) Assumption Two – The decay process operated within a *closed* system so that neither the parent uranium, nor the daughter lead, has either increased or decreased since the time of their origin.

c) Assumption Three – The decay rates have *remained constant* since the beginning of the rock's existence.

Let us consider each of these factors for a moment.

Assumption One is without valid foundation. Why should it be assumed that all of the lead within a sample is the result of a decay process? Why could not lead have been formed within the sample at the time of its initial creation? There is absolutely no valid reason for assuming that such could not have occurred – unless one is attempting to "rig" a dating system.

Henry Morris addresses this very matter. He notes that one of the main problems with the dating processes

> "is the assumption that the amount of radiogenic daughter element present – lead, argon, or strontium [depending upon the particular process being employed] – has all been formed by radiometric decay from the parent element – uranium, potassium, or rubidium. The probability is strong, however, that all these radiogenic 'daughter' isotopes were either formed *in situ* with their 'parents' at the time of creation, or else incorporated with them at the time of magma emplacement, so that the 'apparent ages' were built into the radioactive minerals right from the start when they were formed."[9]

Even evolutionary scientists concede this possibility. Seidemann has noted:

> "potassium-argon dates of these rocks may be subject to inaccuracies as the result of seawater alteration. Inaccuracies may also result from the presence of excess radiogenic Argon-40 trapped in rapidly cooled rocks *at the time of their formation*" [emp. added].[10]

And so, the first assumption is highly suspect.

Assumption Two is significantly debatable as well.

There is an increasing body of evidence that indicates that both parent and daughter elements, under the proper conditions, can migrate in the rocks.

Again we quote from Dr. Morris:

> "... a more common and more likely source of error in radiometric ages arises from the closed system assumption, an assumption which could almost never be really valid. These radioactive methods are always applied only in igneous rocks, and these have all been affected by numerous tectonic, metamorphic, and hydrologic forces. It is almost inconceivable that any mineral could remain a closed system for a billion years of fracturing, folding, solvent action, and other such phenomena. Geochronologists recognize this to be a serious and common problem."[9]

Assumption Three is equally without scientific support, and is, in fact, contradicted by known data. For example, recent research has shown that while decay rates appear to be constant within narrow limits, under certain circumstances they may be altered considerably.

This was forcefully observed in an article by Frederic B. Jueneman in an issue of **Industrial Research and Development**, Dr. Jueneman declared:

> "The age of our globe is thought to be some 4.5 billion years, based on radiodecay rates of uranium and

thorium. Such 'confirmation' may be short-lived, as nature is not to be discovered quite so easily. There has been in recent years the horrible realization that radiodecay rates are not as constant as previously thought, nor are they immune to environmental influences.

> "And this could mean that the atomic clocks are reset during some global disaster, and events which brought the Mesozoic [Age – of the geologic time scale] to a close may not be 65 million years ago but, rather, within the age and memory of man."[12]

This is an utterly devastating admission. It concedes that if some global disaster has occurred in the ancient past (what about the biblical Flood?), then the radiometric clocks are virtually worthless. Dr. Jueneman certainly has not endeared himself to his evolutionary colleagues by this honest confession.

Other writers have spoken to this issue as well. For example, Dr. William S. Beck, an ardent evolutionist, has stated:

> "When all is said and done, there seems to be evidence that even the 'laws of nature' are changing. Modem physics suggests the possibility that changes are taking place in the speed of light and *the rates of chemical reactions*. In other words, the universe is changing, and it becomes hazardous to attempt calculations concerning the very remote past and the future" [emp. added].[13]

Dr. Kenneth L. Currie, of the Canadian Geological Survey, comments:

> "A century of experience and experiment has shown that the selection and calibration of a geological clock is a task full of difficulty and hidden pitfalls. Natural processes in general *do not act at fixed rates.* The assumption that an average rate taken over a long period of time can be extrapolated is generally unsatisfactory."[14]

One of England's scientists, who endorses the idea that the earth is billions of years old, summarizes this matter rather well when he acknowledges: "Perhaps the most questionable parts of geology are the dates, which are bound to be lacking in absolute proof."[15]

The information introduced above surely ought to demonstrate that the evolutionary wave-of-the-hand assertions regarding the alleged great anitquity of the earth are simply a tactic designed to buttress the indefensible theory of Darwinistic infidelity. The Christian need not be intimidated by the chronological propaganda of those who have a personal interest in rejecting the clear testimony of Holy Scriptures regarding the history of this planet.

# 4 Evolutionary Clocks: In Need of Repair

There are numerous evidences which clearly reveal that the evolutionary dating methods are not reliable. First of all, it is well-known that frequently a rock sample will be tested by several different radiometric dating techniques, and yet, widely divergent results will be yielded.

For example, in the Grand Canyon there is a lava formation, known as the Cardenas Lavas, that has been tested by a couple of different radiometric methods. Some scientists, for example McKee and Noble, by the use of the potassium-Argon method determined, that the flow could be as young as 860 million years.

When, however, these same men employed the Rubidium-Strontium technique, it was indicated that the samples could be at least 1,160 million years old.[15] This is a discrepancy of some 300 million years – hardly an example of an accurate time-keeping mechanism.

In Nigeria, the same basalt rock was "dated" by

four techniques and the results were as follows:[16]

    a) Conventional Geology – Upper Tertiary, c. 2-26 million years

    b) Fission Tracks – Less than 30 million years

    c) Potassium-argon – 95 million years

    d) Uranium-helium – 750 million years

Many similar discrepancies in the comparative time-clocks could be demonstrated.

Evolutionists have a way of dealing with "dates" that do not fit their *desired goal,* however. When a sample yields an age which does not square with an anticipated result, it simply is dismissed as inaccurate. Note this startling admission from Hayatsu:

> "In conventional interpretation of potassium-argon age data, *it is common to discard ages* which are substantially too high or too low compared with the rest of the group or with other available data *such as the geologic time scale.* The discrepancies between the rejected and the accepted are *arbitrarily* attributed to excess or loss of argon."[17]

It is crucial to observe that adjustments are made to fit with the *geologic time scale.* As we have shown elsewhere, the geologic time scale is a fabricated, paper chart specifically arranged for the sole purpose of providing support for the theory of evolution. It has no basis in reality.[18]

There is no better way to demonstrate the unreliability of the radiometric clock system than by cataloging a number of the monumental dating blunders that have woefully embarrassed the disciples of Darwin. Consider the following.

1) Studies on submarine basaltic rocks from Hawaii, known to have formed *less than two hundred years ago,* when dated by the potassium-argon method, yielded ages from 160 million to almost 3 billion years.[19] That's like a lady who weighs 100 pounds stepping on a set of scales and seeing the dial register 1.5 billion pounds, or 750,000 tons! I think she might reasonably assume that something is drastically wrong with her scales.

2) "Studies on recent volcanic rocks have been reported by a British consulting engineer, and he finds that there are serious discrepancies in their 'age-dating' by conventional methods. Research on ten samples from

Azores, Tristan de Cunha, and Vesuvius, of rocks *known to be very young*, give 'ages' all the way from 100 million to 10.5 billion years" [emp. added].[20]

A technique that is employed to date objects that once were alive is known as the Carbon-14 method. This system was developed by W. F. Libby of the Institute of Nuclear Studies and the University of Chicago, between 1945-1959. Since this technique was invented, tests have been done on thousands of organic objects, e.g., wood, bone, charcoal, etc. Contrary to the impressions of the average layman, this method, as with those previously discussed, is also fraught with assumptions that are highly questionable. A discussion of these may be read in Henry Morris' book, **Scientific Creationism** (San Diego, CA: Creation Life Publishers, 1974), pp. 161ff.

Even Libby, who was awarded the Nobel Prize for his work in this area, was quite candid regarding the limitations of the process. He wrote:

"You read books and find statements that such and such a society or archaeological site is 20,000 years old. We learned rather abruptly that these numbers, these ancient ages, are not known; in fact, it is at about the time of the first dynasty in Egypt that the last historical date of any real certainty has been established."[21]

Froelich Rainey wrote: "Many archaeologists still think of radiocarbon dating as a scientific technique that must be either right or wrong. Would that it were so simple!" He went on to assert that 1870 B.C. (plus or minus 6) is "the earliest actual recorded date in human history."[22] It is generally conceded that dates yielded by the C14 method that are older than 2,000 to 3,000 years are not to be regarded as reliable.

Again, though, the capriciousness of this technique is revealed by the following examples.

1) The shells of *living* mollusks, when tested by the C-14 process, revealed ages of up to 2,300 years.[23] That is what one might call a "Methuselah" mollusk!

2) Freshly-killed seals have been dated at 1,300 years, and mummified seals, dead only some thirty years, have yielded ages as high as 4,600 years.[24]

3) "Muscle tissue from beneath the scalp of a mummified musk ox found in frozen muck at Fairbanks Creek, Alaska, has a radiocarbon age of 24,000, while the radiocarbon age of hair from a hind limb of the carcass is 7,200."[25] Did the ox wander around bald for some 16,800 years – then finally grew hair.

4) Wood from growing trees has been dated by the C-14 method with the suggested age of 10,000 years.[26] This is strikingly odd since the oldest known tree on earth is the bristlecone pine. One of these, in the Snake Range of Nevada, has been estimated (by its rings) to be some 4,900 years old, but even this may not be accurate for sometimes trees are known to grow more than one ring per year.[27]

In conclusion we must reemphasize, *there is absolutely no genuine scientific evidence that proves the earth is billions of years old.* The dating methods are built upon uniformitarian (the present is the key to the past) assumptions which are an intrinsic part of the evolutionary scenario.

# 5 The Bible and Inspired Chronology

For the Christian, the Bible is the final authority on any subject that it specifically addresses. One must be sure, of course, that he is responsibly reading and correctly understanding the Holy Scriptures.

Since the Bible was given, however, for the purpose of human enlightenment and acceptation, it is expected that the average person ought to be able to read the clear testimony of the Word, and, depending upon the integrity of his disposition, either accept or reject its message.

We do not for one moment subscribe to the modernistic notion that the Bible is *partially* inspired, i.e., that its exalted religious truths are from God, but that its allusions to science or history are suspect. (For a discussion of the inspiration of the Bible, see our book, **Fortify Your Faith**; see also our volume, **The Bible & Science**). The Scriptures are not designed to be a technical textbook upon historical, geographical, scientific, etc., matters, but whenever these areas are incidentally addressed, one may be sure that the inspired documents are completely reliable. The sum of God's word is truth (Psalms 119:160 ASV).

Frequently we are told that the Bible is not concerned with chronological matters. With reference to the Genesis record, it is asserted, Scripture is directed toward the *Who* and the *why*, and not with the *how*, or the *when* of earth's creation and its inhabitants.

Whenever I hear a statement of that nature, an alarm goes off in my mind. I know that I am dealing with a misguided religionist who has little regard for the authority of the ancient text; or else, I am hearing the naive parroting of some novice who gullibly has adopted the presuppositions of evolutionary dogma – without being aware of the source whence they derive.

The following is a typical example of the type of attitude of which we speak.

> "A reading of the first few chapters of Genesis leaves one with the very definite general impression that life has existed on earth for, at the most, a few thousand years. That conclusion is in conflict with the conclusions of modem science [evolutionism] that the earth is ancient. However, nowhere does a Biblical writer give us an age for the earth or an age for life on earth .... Inasmuch as Scripture does not state how old the earth is or how long life has existed on earth, one is free to accept, if he wishes, the conclusions of science [evolutionary chronology]."[28]

The truth of the matter is, the Bible, being a book grounded in history, is utterly filled with chronological data. Professor Edwin Thiele has well noted:

> "Chronology is important. Without chronology, it is not possible to understand history, for chronology is the backbone of history. We know that God regards chronology as important, for He has put so much of it into His Word. We find chronology not only in the historical books of the Bible, but also in the prophetic books, in the Gospels, and in the writings of Paul."[29]

Does the Bible speak, in any sense, concerning the age of humanity, or that of the earth? Indeed it does.

We are not suggesting that one can settle on a specific date for Genesis 1, as did John Lightfoot (1602-1675), the famous Hebraist of Cambridge, who contended the creation occurred during the week of October 18 to 24, 4004 B.C., and that Adam was made on October 23rd at 9:00 A.M. forty-fifth meridian time!

We are definitely contending, though, that the Bible gives a chronological framework that generally *limits* the age of humanity, and, correspondingly, the age of our planet. Therefore, the antiquity of the earth must be confined to a span of several thousand years – certainly not millions or billions. This reality is a deadly argument against the theory of evolution, and the Christian must never surrender it.

Our argument will be developed syllogistically as follows.

First, if it is the case that the Bible limits the longevity of humanity to within several thousand years, and, if it is likewise the case that Scripture indicates that mankind and the earth are dated at the same approximate time, then it necessarily follows that the age of the earth is to be measured in terms of several thousand years.

Second, it *is* the case that the Bible limits the longevity of humanity to within several thousand years, and, it is likewise a fact that mankind and the earth are dated at the same approximate time.

Third, it is, therefore, the case that the earth is to

be measured in terms of several thousand years, not billions.

The argument is set forth in a valid form. Our responsibility now will be to argue our case. If the minor premise can be scripturally demonstrated, the case for a young earth will be proved, and thus, *by the concession of all, evolution is discredited.*

I am aware of the fact that this argument will carry no force whatever with those who reject the inspired authority of the Holy Scriptures. I am not addressing infidels with this logic. I am assuming that I am reasoning with those who believe that the Bible is the inspired Word of God and that they, with reasonably intelligent minds, can understand what the Lord has said in his sacred book.

With this foundation laid, we will proceed to establish our case as per the previously stated syllogistic argument.

# 6 Biblical Genealogies and Human History

Is it true, based upon biblical evidence, that human history must be limited to within several thousand years? We confidently affirm this to be the case. The proof is to be found in the inspired *genealogies* recorded in the divine Scriptures.

The study of scriptural genealogies is not a worthless endeavor, as some believe. These records play a vital role in biblical literature, as evidenced by the fact that they occupy so much space in the sacred narrative. Though genealogies (and chronologies) served various functions in the literature of Scripture, one of their main purposes was to show the *historical connection of great men to the unfolding of Jehovah's redemptive plan*. These lists, therefore, form a connecting link from the earliest days of humanity, to the completion of Heaven's salvation system. That these genealogical and chronological lists must be *substantially complete* to have any evidential value should be fairly clear.

For example, the inspired writer of Hebrews, in contending for the heavenly nature of Christ's priesthood, argued that the Savior could not have functioned as a priest while he was living upon the earth since God had a levitical priesthood to accomplish that need (cf. Hebrews 8:4). Jesus Christ did not qualify for the levitical priesthood for "it is evident that our Lord hath sprung out of Judah" (Hebrews 7:14). How could it have been "evident" that Jesus Christ was from the tribe of Judah unless there were accurate genealogical records by which such a statement could be checked? The argument assumes that the readers of this epistle would not dispute the ancestry of Christ in view of the reliable Jewish documentation available.

Some contend, however, that the genealogical lists in the Bible are of no value in determining the length of human history. It is alleged that there is no necessary conflict between the Old Testament view of man's origin and the modem anthropological theory that human history reaches back several million years. One author states:

"Any attempt to ascribe a specific *or even a general*

*age* to either man or the Earth from a Biblical standpoint is a grievous error"[30] (emp. added).

Again: "The time of man's beginning is not even hinted at in the Bible. There is no possible way of determining when Adam was created."[31]

The same writer then discusses the so-called geologic time table, without any indication that he disagrees with its time scale, which, according to evolutionists, chronicles earth's history back some 4.5 to 5 billion years. (For a study of the geologic time scale, see our book, **The Mythology of Modern Geology**.) The writer then concludes: "If geologists are correct in their dating methods, man is a very recent newcomer to this planet."

It might be well to remind ourselves at this point that evolutionary anthropologists contend that true man (*Homo sapiens*) has been around for approximately 3.6 million years (though such dates, like airline schedules, are subject to change at a moment's notice!). In fact, as we noted earlier, some are now contending that man has been on earth for possibly 6 million years. We ardently affirm that the foregoing assertions are patently false, and that they represent a deplorable compromise of the biblical position regarding man's antiquity.

The Bible unquestionably teaches that Adam was "the first man" (1 Corinthians 15:45). Now in Luke 3:23-38, the inspired historian traces the human lineage of Jesus Christ (who lived approximately two thousand years ago) all the way back to Adam. If it is true that the Scriptures give no hint as to when Adam was created, and if it is likewise true that it is possible to harmonize the biblical record with the assertions of modern anthropology, then it is clear that several million years must somehow be squeezed into the genealogical record of Luke's gospel account.

## Genealogical Gaps

Every diligent Bible student is well aware of the fact that the biblical genealogies do not always reflect strict father-son relationships. There is no question but that these lineage records do occasionally contain some gaps; but here is the real question: *are these gaps sufficient to accommodate millions* of years? Note the following examples.

First, in Ezra 7:3-4, from Zerahiah to Amariah involves only four names, while in 1 Chronicles 6:6-10, from Zerahiah to Amariah there is a total of ten names. There is thus a gap of six names in Ezra's list. This gap of a half-dozen generations obviously involves but a few years relatively speaking; certainly not hundreds of thousands of years.

Second, an abbreviated genealogy from Levi to Moses mentions only four names (Exodus 6:16-20), while a lineage of the same general time span, from Joseph to Joshua (1 Chronicles 7:20-27) involves eleven names. There could thus be some seven generations deleted between Kohath and Amram in Exodus, but this gap omits only about three hundred years altogether, not thousands of years. (It might be pointed out, however, that C.F. Keil, in his work **The Pentateuch**, I, argues for the possibility of two men named Amram in Exodus 6:18-20.)

Third, in 1 Chronicles 26:24, Shebuel, "a ruler over the treasures" during the administration of David, is said to be the "son of Gershom, the son of Moses." The term "son," in one instance here, is used in the sense of *descendant*, because about four hundred years separate Gershom from David's time. Again, though, this has no *geological* time significance whatever.

## The Genealogies of Christ

There are two genealogies of Jesus in the New Testament, Matthew 1:2-16 and Luke 3:23-38. Matthew's account, designed for Jews, was intended to establish Christ as a descendant of Abraham and David (1:1), and thus the illustrious *seed* promised to those Old Testament worthies. Matthew's genealogy is artistically arranged into three sets of fourteen generations (perhaps to facilitate memorization), hence, between Joram and Uzziah (1:8) he omits three names. There may be other names omitted as well, since Matthew lists fourteen generations between David and the Babylonian Captivity, and fourteen generations between the Captivity and the birth of Christ, while in Luke's list, these same two spans chronicle twenty or twenty-one generations respectively. Luke's record, written especially for the Gentiles to emphasize the Lord's solidarity with mankind, is clearly more exhaustive.

We must point out, however, that the reason we know that these gaps exist is because *the missing names are supplied elsewhere in the Bible*. Really, therefore, there are no provable missing links when the total biblical picture is compiled.

Even though there are plainly some gaps in some of the genealogical accounts, as we have noted already, they are relatively minor. And *this can be demonstrated.*

According to Luke's record, there are fifty-five generations from Abraham to Joseph (the "supposed" father of Jesus). The account in Luke is most likely the genealogy of Jesus through Mary. Now archaeologists have established that the era between Abraham and Jesus covers, at the most, some two thousand years.[32] Simple mathematics will reveal that fifty-five generations spanning two thousand years allows for an average of only about *forty years* per generation – including any possible gaps. There simply cannot be, therefore, any *huge* gaps from Abraham to Jesus.

Furthermore, Luke's list mentions only twenty names from Abraham back to Adam (a number of whom were renowned for their longevity). The sacred historian even includes Cainan between Arphaxad and Shelah (3:35-36) – a fact not mentioned in the genealogies of the Hebrew Old Testament (though included in the Septuagint). Even if we grant the possibility of some omissions in this record (and that cannot be proved), why should one assume that a drastically *different type* of genealogy is employed from Adam to Abraham, than is utilized from Abraham to Jesus?

In other words, does it seem sensible that the slightly more than fifty generations from Abraham to Christ bridge only about two thousand years; and yet, allegedly, some 3 million years (or more) can be pressed into the twenty generations preceding Abraham? Is this a reasonable approach to Bible interpretation? Assuredly it is not; *and no one would dream of it if not for the fact that they were attempting to stretch the genealogy to accommodate evolutionary chronology!*

If it is the case that the genealogy from Adam to Abraham embraces several million years, it is the epitome of irrelevancy, and a rational attempt to interpret the Scriptures is hopeless!

While allowing for minor elasticity within the biblical lineages, J. Barton Payne declared that drastic links leave "the Bible's detailed lists of figures as generally pointless and also posits an unusually high proportion of omitted links."[33]

In discussing the purpose of the scriptural genealogies, Dr. John Klotz cautions against trying to construct a strict chronological calendar to determine the *exact* time of creation. He nevertheless concedes: "God apparently did want to show us that the earth is not billions of years old."[34]

A portion of the major premise of our argument, as set forth at the end of the previous chapter, was this: *the longevity of humanity must be limited to within a few thousand years.* We believe, therefore, on the basis of the clear testimony of Scripture, a consistent method of interpretation, and archaeological evidence, that that element of the premise has been established beyond doubt.

It will now be our chore to argue that the Bible places the creation of man, and the origin of the earth, at the same general time, and, therefore, that the earth must be likewise dated in terms of several thousand years, as opposed to billions of years.

# 7 Man: "Johnny-come-lately" or "From the Beginning"?

Is it true, as evolutionists allege (and some religionists agree) that man is really just a Johnny-come-lately upon this planet? Does the Bible supply any information concerning the *relative* ages of human-kind and the earth?

Let us rehearse the evolutionary scheme of things at this point. It is contended that our earth is approximately 4.5 to 5 billion years old, but that man appeared a scant 3.6 million years ago. As we have earlier noted, recent estimates (still highly disputed) are pushing this back even more remotely. If these figures were to be accepted as approximately accurate, then simple mathematics would reveal that man is but about 1/1250th of the age of the earth. If such were true, humanity is but a speck on the panorama of history. Perhaps the following illustration will dramatize the force of this.

Suppose we let one day represent the sum of earth's history (as seen from the evolutionary viewpoint). This means that the supposed 4.5 billion years are represented by the 86,400 seconds of a single day. Since humanity's age is assumed to be only 1/1250th of the earth's, man, on this one-day time scale, would be slightly more than *one minute and nine seconds old!*

Look at it another way. If one were to draw a horizontal line *one hundred feet* long, and then, at the right end, directly underneath, he drew another line only *one inch* long, he could vividly see the difference between the alleged respective ages of the earth and man, from the *evolutionary* perspective.

Accordingly, if the whole of earth's history were viewed from our present vantage point, human existence commenced virtually at the *end* thereof. This point needs to be emphasized forcefully. The evolutionary theory, along with the compromising views that accommodate it, do not allow that man originated at the *commencement* of creation history. Anyone, therefore, who accepts the evolutionary chronology of geo-human history, cannot possibly believe that humanity has been on this planet from *the beginning of the creation.* Yet, this is what the Bible affirms – *repeatedly*.

## The Beginning

The biblical expression "in the beginning," "from the beginning," or "since the beginning" denotes the originating point within a certain time frame – the precise frame being determined by the context. In the Scriptures, over and over again, human-kind is represented as existing since *the beginning* of the original creation.

While it is true that the expression can involve a minor degree of relativity, such obviously is quite limited; otherwise, language is meaningless. In other words, when something is said to be "from the beginning" of a certain period, there must be reasonable proximity involved. With this in view, let us consider the following Bible evidence.

## The creation account

Genesis 1:1 declares: "In the beginning God created the heavens and the earth." Subsequent to that statement, the events of the creation week are outlined. On the sixth day, God said: "Let us make man in our image, after our likeness" (v. 26).

Any reasonable reading of this majestic context can lead only to the conclusion that man was brought into existence during the same week the earth was created, and all exegetical manipulative methods (e.g., the gap theory, day-age theory, etc.), which have sought to negate this view, have proven absolutely futile.

## The sabbath day

In Exodus 20:8ff, the Hebrew people were charged with the responsibility to: "Remember the sabbath day, to keep it holy." Hence, they were to do all of their work in six days, but the seventh day was to be dedicated unto Jehovah. In commenting upon this command, Moses gives the historical basis for it. "For in six days Jehovah made heaven and earth, the sea, and all that in them is..." (v. 11; cf. 31:17).

Here, it is clearly affirmed that the entire creation, including mankind, was brought into existence during the same week as the earth. It is also important to note that the context defines the nature of the "days" of that initial week – they were the *same type* of days as the sabbath day, i.e., a literal day, and not an eon of time.

## Wisdom's fellowship

In Proverbs 8:29-31, the inspired writer, personifying wisdom, declared that *when* God "... marked out the foundations of the earth: *then* I [wisdom] was with him ... rejoicing in his habitable earth; and my delight was with the sons of men." As one learns in exploring grammar, a "when ... then" clause is employed to denote contemporaneous time frames – as in *"when* the clock strikes, *then* it will be precisely midnight." So, in this passage, the writer clearly sees *the founding of the earth,* and wisdom's fellowship with *people,* as being in close proximity.

## The foundations of the earth

When Isaiah was contrasting the greatness of Jehovah with the impotence of idols, he asked: "Have you not known? Have you not heard? Has it not been told you *from the beginning?* Have you not understood from *the foundations of the earth?"* (Isaiah 40:21).

Note how the prophet parallels the expressions "from the beginning" and "from the foundations of the earth," and suggests that man had known of God's nature since that time. Are these statements to be expunged from the Bible simply on the basis of evolutionary chronology?

## Things hidden

During his earthly ministry, Jesus announced: "I will utter things hidden from the foundation of the world" (Matthew 13:35). From whom had these new truths been hidden "from the foundation of the world"? Would that statement make any sense if no rational human beings existed until billions of years *after* the foundation of the world?

## From the beginning, he created them male and female

Christ affirmed: "But from the beginning of the creation, Male and female made he them" (Mark 10:6). This clearly dates the first human couple from the creation week. "Beginning" *(arche)* here is used to mean "absolute, denoting the beginning of the world and of its history, the beginning of creation." And "creation" *(ktiseos)* involves the "sum total of what God has created."[35] Bloomfield noted that "creation" in Mark 10:6 "signifies 'the things created,' the world or universe."[36] Unquestionably, this language puts human-kind back at the very dawn of time.

To reject this clear truth, one must either contend that: (a) Christ knew that the universe was in existence billions of years prior to man, but, accommodating himself to the ignorance of his generation, *deliberately misrepresented* the situation; or, (b) The Lord, living in pre-scientific times, was ignorant of such matters (despite the fact that he was at the beginning as Creator – John 1:3; Colossians 1:16). Either of these allegations is a reflection upon the divine nature of Jesus Christ, and is blasphemous.

## Suffering since the creation

In the Savior's Olivet discourse, he prophesied of the great suffering that would descend upon Jerusalem in connection with her destruction (70 A.D.). Christ announced: "For those days shall be tribulation, such as there hath not been the like from the beginning of the creation which God created" (Mark 13:19). "Tribulation" is a form of distress to which human beings sometimes are heir; it is *always* used in the Bible of *people*. The Savior's warning here contains the implication that affliction has been the lot of humanity *from the beginning of the creation*. Human suffering began with the apostasy in Eden, and that sad scene was sufficiently near the creation week to justify the language employed by Jesus.

## Prophets persecuted from the foundation of the world

In Luke 11:45-52, Christ rebuked the rebellious Jews of his day and foretold the horrible destruction that soon would be visited upon them. He charged them with following in the steps of their ancestors, hence announced that upon them would come "the blood of all the prophets, which was shed from the foundation of the world" (50). Then, with parallelism characteristic of the Hebrew mode of expression, the Lord re-phrased the thought by saying, "from the blood of Abel unto the blood of Zachariah...."

Now here is a very important point. Jesus places

the murder of Abel back near the *foundation of the world*. Of course that first murder occurred a few years after the initial creation week, but it was close enough, from a first-century viewpoint, to be associated with the beginning of the world. However, if the world came into existence several billion years before the first family, how could the shedding of human blood possibly be declared to extend back to the foundation of the world? That would be some hyperbole!

## The devil, murderer from the beginning

Though Genesis does not declare how long Adam and Eve were in the garden of Eden prior to the event of their fall, it obviously was not very long. This is revealed by the fact that Christ, referring to the curse of death upon the human family (cf. Romans 5:12), said that the devil "was a murderer from the beginning" (John 8:44). Again, the implication is clear: humankind extends back to the beginning.

## The invisible things of him

Paul's letter to the Romans contains the following

dramatic statement:

> "For the invisible things of him since the creation of the world are clearly seen, being perceived through the things that are made, even his everlasting power and divinity; that they may be without excuse…" (1:20).

The phrase "since the creation of the world" translates the Greek *apo ktiseos kosmou*. The preposition *apo* is used: "To denote the point from which something begins,"[37] and thus is rendered "since." The term for "world" is *kosmos*, and it refers to "the orderly universe."[38] *Kosmos* primarily meant order or arrangement, but came to be applied to the universe due to the order characteristic of it.[39] Trench observes that *kosmos is* "the material universe ... in which man lives and moves, which exists for him and of which he constitutes the moral centre."[40]

So Paul declares that the orderly universe, since the time of its commencement, has testified of the invisible things of God, through the things that are clearly seen and perceived by man. The term "perceived" is from *noeo,* a word used for rational intelligence, while "clearly seen" is an intensified form of *horao*, a word which "gives prominence to the discerning mind."[41] Both "perceived" and "clearly seen" are present tense forms, and they denote "the continued manifestation of the being and perfections of God, by the works of creation from the beginning...."[42]

The apostle's point is perfectly clear: the power and divinity of God, as revealed in the things that are made, have been observable to human intelligence since the creation of the world. Man thus has existed from the beginning. The earth is not billions of years older than mankind.

## The works of creation finished

The author of the book of Hebrews, after citing Genesis 2:2, "God rested on the seventh day from all his works," further declared that "the works [of creation] *were finished from the foundation of the world*" (4:3-4). Was man one of those works? Indeed. He, therefore, has existed "from the foundation of the world." Thus, both Old and New Testaments affirm that Jehovah concluded his creative activity at the end of the sixth day.

How does this harmonize with the notion that man is a "new-comer" to this planet? If man, a uniquely new type of creature, did not arrive until billions of years after the original creation, Genesis 2:1-2, and Hebrews 3:3-4 convey a false impression.

## Sacrifices needed since the creation of the world

In Hebrews 9, the inspired writer drew a vivid contrast between the repetitious nature of the sacrificial system of the Mosaic economy, and the offering of the Lord's body "once" (i.e., one time) for sin (cf. 24ff). In that connection the author asserts that if Jesus' death had been like that of the Old Testament regime, he would have suffered often "from the foundation of the world." The inference is crystal clear. A sacrificial offering of some type had been needed *since the foundation of the world.*

That being the case, it is obvious that *sin* had been present virtually since the dawn of time. And inasmuch as sin is an act of rebellious man, it is clear that human history extended back to the foundation of the world. How much more evidence would it take to establish the fact that the Bible places humanity back at the very commencement of creation?

## Common understanding in the first century

Finally, it is very apparent that the Jews of the first century believed that the antiquity of mankind extended back to the creation (an impression *nowhere* contradicted by an inspired person). For instance, certain mockers asked regarding Christ: "Where is the promise of his coming? For, from the day that the fathers fell asleep, all things continue as they were from the beginning of the creation" (2 Peter 3:4). These citizens of the apostolic age took it for granted that the patriarchs of old existed from the time of the creation.

Furthermore, the blind man healed by Christ, in expressing his amazement at the miraculous powers of the Savior, exclaimed that, "Since the world began it was never heard that any one opened the eyes of a man born blind" (John 9:32).

While these final two points would not be sufficient alone to make the case, they do nevertheless reflect a harmony between divine teaching (as cited above), and first-century Jewish opinion regarding man's history.

We might observe that the Christian/Jewish view of earth history stood in marked contrast to the pagan concepts of antiquity. The prevailing view among the ancient Greeks, Egyptians, Babylonians, etc., was that the earth was very old – probably eternal.

In view of the overwhelming nature of this testimony concerning the contemporaneous nature of human and earth history, the second part of the minor premise of our original argument stands established. Human history and earth history started at substantially the same time – only six days apart

And since, therefore, human history can only be measured in terms of several thousand years, earth history likewise must be limited to a few thousand years. This being the case, evolution can not have occurred – no matter what methodology is surmised! Remember, "time is the hero of the plot." In this instance, the "hero" has become the villain!

# 8 Theories That Torture the Scriptures

T. H. Huxley (1825-1895), the radical evolutionist who was known as "Darwin's bulldog," once spoke of certain religious writers who "torture texts to make them confess the creed of science." He was exactly right. There are numerous people, who identify themselves as Christians, who have been so thoroughly intimidated by the assertions of evolutionary scientists that they will resort to almost any type of textual distortion in order to force the Scriptures into harmony with modern scientism.

In this brief study we will mention several such efforts. They are: (1) Theistic Evolution; (2) the Gap Theory; (3) the Day-Age concept; and (4) Progressive Creationism.

## Theistic Evolution

Theistic Evolution is the mongrel notion that suggests a supreme Being was responsible for the creation of matter, and possibly the origin of an initial life-force. Perhaps He dipped His finger into the developmental process a few times along the way, but for the most part, the world of living creatures is the result of the evolutionary process.

Theistic evolutionists are converts of Charles Darwin, yet they cannot shake the notion that God somehow had something to do with the proliferation of living creatures. Theistic evolutionists come in varieties and degrees; some adopt more of the Darwinian philosophy than others; all of them have surrendered significant truth.

The fact is, there are numerous points of disagreement between the evolutionary scenario and the Genesis record. For example, evolution alleges that our planet's first surface was land, water later seeping from its interior. Genesis states that the first stage of earth's composition was water, and dry land appeared subsequently (1:2, 6, 9). Moses declared that the globe's first life forms were plants (Genesis 1:11); evolution contends that marine life came before plants. Evolution alleges that reptiles preceded birds;

the Scriptures assert that birds came before "creeping things" (Genesis 1:21, 24). The Bible teaches that God created human beings as man and woman (Genesis 1:27; 2:7, 22; cf. Matthew 19:4); Darwinism suggests that a bi-sexual blob evolved into distinct sexes, that, via some ape-like ancestor, ultimately became human. There simply is no way to bring Moses and Darwin together without the most exaggerated form of distortion of the sacred text.

## The Gap Theory

The Gap Theory was first proposed in 1814 by Thomas Chalmers of Edinburgh University. It alleges there was a vast span of billions of years in between Genesis 1:1 and 1:2. Professor Oswald Allis pointed out that this idea "gained favor about a century ago as a means of bringing Genesis 1 into harmony with the findings of geologists."[43] The Gap Theory has been popularized by a footnote in the **Scofield Reference Bible**.

The theory argues that during this gap between Genesis 1:1 and 1:2, there lived successive generations of plants, animals, and perhaps even a pre-Adam race of men. Gleason Archer, of Fuller Theological Seminary, identified certain of those ancient fossil "men" (e.g., Pithecanthropus, Swanscombe, Neanderthal, etc.) with a pre-Adam race. According to some that hold to this view, God destroyed the original creation because of a Satanic rebellion. It is suggested, therefore, that Genesis 1:2 should be rendered, "the earth *became* waste and void."

This question is certainly appropriate: Does it seem reasonable that an original creation and subsequent catastrophe, which would dwarf every other catastrophic event in Scripture, should be totally ignored elsewhere in the entire Bible? There is no evidence in the Scriptures for this fanciful idea, and, as a matter of fact, it contradicts the sacred narrative in several important particulars. Consider these points.

First, the conjunction "and," which introduces verse 2 does not convey the impression of a vast gap of time between these passages.

Second, there is no reason within this context to translate "was" (1:2) by "became." The Hebrew verb is *hayah*. It is found 1,522 times in the Old Testament, and though it may occasionally be rendered "became" (twenty-two times), the context must require it. And no such requirement is here indicated.

Several years ago, twenty leading scholars of the Hebrew language in the United States were asked if there was exegetical evidence for a gap between Genesis 1:1 and 1:2. They emphatically replied, "No."[44]

Stigers has observed:

> "The cataclysmic theory (also called the restitution theory) respecting v. 2 can have no place in a proper translation. The construction of 'became void,' etc., is not justified by Hebrew syntax. When the verb 'to be' (*hayah*), *is* to be construed as 'became,' the addition of the prepositional *lamedh* is required with the following word to provide this meaning, and this preposition is absent here."[45]

Third, Adam was given dominion over *"every living thing* that moveth upon the earth" (Genesis 1:28). It is impossible to harmonize this statement with the notion that millions of living creatures had lived already and died (many having become extinct) by this time.

Fourth, at the conclusion of the sixth day, "God saw every thing that he had made, and, behold, it was very good" (Genesis 1:31). If Jehovah's original creation had become contaminated through Satan's rebellion, and thus was destroyed subsequently, and the new re-creation rested upon a vast "graveyard" of corruption, it is difficult to see how the expression "very good" would have been appropriate to that situation – even remotely.

Fifth, Adam is called "the first man" (1 Corinthians 15:45), and Eve was designated as "the mother of all living" (Genesis 3:20). These expressions exclude the idea that there was a pre-Adam race.

Weston W. Fields, in his book, **Unformed and Unfilled** (Nutley, NJ: Presbyterian & Reformed, 1976) has utterly demolished the Gap Theory. There is no vestige of biblical support for it.

## The Day-Age Theory

Advocates of this theory argue that the "days" of the creation week were not literal days of approximately twenty-four hours; rather, it is claimed, each day *symbolically* represents millions or billions of years. This viewpoint is also fraught with insuperable difficulties, both biblical and scientific. We will consider but a few of these.

## Normal "days" in the creation

First, the concept is unmindful of several obvious contextual factors which indicate that normal days were under consideration in the creation narrative, such as "the evening and the morning" (1:5, 8, etc.). Has anyone ever seen an "eon" with an evening and a morning? Note the Mosaic distinction between "days" and "years" in 1:14.

> "And God said, Let there be lights in the firmament of heaven to divide the day from the night; and let them be for signs, and for seasons, and for *days* and *years:*"

If "days" actually represent years, what do "years" signify?

## Ordinal numeration used

Second, the Day-Age Theory overlooks the general use of the term "day" as employed in connection with the ordinal numeral, e.g., "first day," "second day," etc. Note the use of the expressions "first day," "second day," etc., in Numbers 7:12, 18ff.

Does the honest student have difficulty in determining the meaning of those expressions in that context? Of course not, because the *evolutionary motive* is not thrust into that setting!

## Moses ignored

Third, this theory ignores Moses' personal commentary on the meaning of the creation "days" in Exodus 20:11, where, as we earlier observed, he equates the days of the creation week with an ordinary sabbath day, such as the Hebrews were required to observe. Let someone try substituting the term "age" for "day" in Exodus 20:10-11, and just see how much sense it makes.

## Evening and morning

Fourth, the days of Genesis 1 are all divided into equal portions of light and darkness. If "eons" are under consideration, how could the vegetation of the third day have survived the millions of years of darkness in those subsequent nights, without the light necessary for photosynthesis (manufacture of food for plants by means of sunlight)?

## Scientific problems

Fifth, the botanical world was born on the third day, yet the kingdom of other living things was not created until the fifth and sixth days. How could those plants that are pollinated exclusively by insects have survived? Clover is pollinated by bees, and the Yucca plant has the Pronuba Moth as its only means of pollination. How did these plants endure for millions of years before there was any method of reproduction to accommodate them?

## Implications of the creation week

Sixth, time is measured in various ways. The earth revolves around the sun each 365 days – the year. It is inclined twenty-three and one-half degrees on its axis, determining the seasons. The moon revolves around the earth each 30 days, marking our month. The day is the result of the earth rotating on its axis every twenty-four hours. Thus, the natural movements of the solar system mark the years, seasons, months, and days. But what determines the week?

Alexander Campbell declared:

> "This is a question of great importance – a question that staggers the boldest of infidels, and the most expert theorists ... we affirm that nothing on earth or in heaven, can be assigned as an argument for the week, aside from the fact that the heavens and the earth were created in six days of twenty-four hours each."[46]

Despite the capitulation of those who declare that the creation-week "days" is an open issue, supposedly with "good arguments" on both sides of the controversy, the Day-Age Theory must be seen for what it is – a pathetic effort to throw sweet morsels to the evolutionists – who could care less about Genesis anyhow.

## Progressive Creationism

Progressive Creationism is the concept that God has periodically intervened in the creative process; in between those activities, however, evolutionary development, over millions of years, has occurred. This is but a slightly subdued form of Theistic Evolution.

According to progressive creationists, Genesis 1 is

not to be viewed as an historical account of what actually happened during the first week of earth's history; rather, it is but a "broad outline" of the eons of creation history. This theory is designed to pay some tribute to the creative activity of Jehovah, and yet, it has been a "reed in the wind" to the pressures of evolutionary geology. It filters the biblical record through the unfounded assumptions of modern scientism, and there is no justification for it.

# 9 Physical Evidences for a Young Earth and Humanity

For those who accept the authority of the Bible, the issue of the relative age of the earth is settled. The earth is to be dated in terms of several thousand years – not billions. If the Scriptures are accurate, and there is not the slightest doubt that they are, then there ought to be evidence in the natural world that lends confirmation to this truth. Is there? Assuredly so. Though we can only briefly touch upon this matter, we believe the reader will be singularly impressed with several historical and scientific evidences for a young earth.

## Ancient Civilizations

Do you know how old you are? "Sure," you reply, "I am twenty years of age." How do you *know* that you are? Someone may have told you, but in reality you are depending upon their word. By your appearance it may be assumed that you are in the vicinity of twenty; still, though, that is not exact. The only way to accurately determine your precise age is by means of a written record. You can document your age by your birth certificate. Similarly, the only sure means we have for determining the age of ancient civilizations is by means of the written records they left behind.

When evolutionists contend that man has been upon the earth for three million years (or longer), and when certain archaeologists allege that ancient civilizations can be dated to 50,000 years B.C., they are *speculating* – pure and simple! All of those ancient dates are based upon the radiometric techniques (discussed in Chapter 3), and these procedures are highly unreliable beyond a few thousand years.

The earliest civilization known to us is the ancient Sumerian nation of Mesopotamia. It may be dated at about 3,500 B.C. The civilization of Egypt goes back to approximately 3,000 B.C., and some (e.g., Donovan Courville and Immanuel Velikovsky) have argued that even these figures are exaggerated and perhaps need to be reduced by as much as eight centuries. Hence, so far as *the actual evidence* indicates, human history does not extend back into the millions of years; rather, it is but a few thousand years old.

## Population Statistics

The world's population has now eclipsed five billion people. How many people this earth could adequately support is not known for certain, but some estimate that the upper limit may be around fifty billion. Based upon what we know about reasonable rates of population growth, however, mankind cannot have been upon this earth for even one million years, let alone the three-million-plus figure (or more) which is being advocated now.

For example, if it is assumed that a generation is approximately thirty-five years, and that a family and its succeeding off-springs had only three children, the present world population would require the passing of only some fifty-two generations, which would extend back only eighteen hundred or so years. Obviously, therefore, our model data are extremely conservative.

With that in mind, i.e., that approximately *fifty-two* generations could account for the present population of the entire world, consider this. If humanity had been around for only one million years (which would involve some 28,600 generations), the present world population should be about 10 to the 5,000th power (that is one followed by five thousand zeros. However, *only about 10 to the 100th power people could be crammed into the entire known universe!* It is obvious, therefore, that humanity could not have been multiplying, even at a very conservative rate, for as long as a million years, let alone 3 million – or six million! Population statistics argue for an infant human race.[47]

## Meteorite Fragments

Billions of meteors enter the earth's atmosphere each year, traveling at the rate of about 100,000 miles per hour. Most of them burn up when they enter the atmosphere (about fifty miles up), but some, known as meteorites, hit the earth's surface. Most meteorites are fairly small, but some rather large ones have been discovered. One meteorite, in the American Museum of Natural History in New York, weighs thirty-five tons, and another in Africa weighs some sixty tons.

There is no reason not to assume that this phenomenon is coextensive with the entire span of earth history. If, therefore, the various strata of the earth required billions of years to build, there ought to be evidence of meteorite bombardment down through the various layers of sedimentary (water-laid) strata. Such, however, is not the case. Meteorites have been found only in the *surface,* or in the "younger" sedimentary layers. None has been found in the deeper strata.[48]

This is consistent with the idea that the earth is young and that the sedimentary strata were generally laid down rather rapidly (e.g., at the time of the Genesis Flood). But this circumstance is inconsistent with the uniformitarian concept of a slowly formed earth, built-up over billions of years.

## Decay of Earth's Magnetic Field

One of the important force fields associated with our earth is its *magnetic* field. This magnetic field is due to an intense electrical current, billions of amperes, circulating in the core of our planet. Dr. Thomas G. Barnes, Professor-Emeritus of Physics at the University of Texas at El Paso, who has done extensive research in this area for many years, declares: "It is known that the earth's magnetic field is decaying faster than any other worldwide geophysical phenomenon."[49]

For more than a century and a half, at intervals of every ten or fifteen years, this process has been under investigation. It is known, for example, that only 1,400 years ago, the magnetic field was twice the strength of its present state. If one extrapolates back 4.5 billion years ago, the magnetic field would have been 200 million times stronger than it now is. "This is stronger than all but the most intense magnetic fields in the universe, those in neutron stars, and is impossible to maintain on earth."[50]

By the use of certain mathematical equations, Dr. Barnes, working backwards in time, has postulated that the earth, at the most, could be no more than 10,000 years old. Evolutionists, recognizing the implications of this phenomenon, speculate that there have been "reversals" in the earth's magnetic field, but there is absolutely no scientific evidence that such have occurred.

## Subterranean Pressure Fields

Deep under the surface of the earth are huge reservoirs of oil and water. Many of these reservoirs are characterized by extremely high-pressure fluids. It is this high pressure that produces the "gushers" with which well-drillers are so familiar. Scientists are aware of the fact that these underground pressures are gradually diminishing (much like air seeping from the tire of an automobile). What intrigues them is this: if this seepage has been going on for eons (and they assume these pools are millions of years old), why hasn't the pressure completely dissipated? It is an acknowledged fact that the rock above these pressure pockets is, even under the strictest conditions, porous enough to allow the pressure to escape within a matter of *only a few thousand years.*

Dr. Melvin Cook, who served as a professor at the University of Utah, and was President of IRECO Chemicals (1968 winner of the Nitro Nobel Award), wrote the following:

> "...[T]he permeability of the rock above the trapped high pressure fluid zone ... is so high even in the tightest

traps that the fluid will leak out of the formation at the prevailing pressures in only a few thousand years."[51]

Again, this is compelling evidence for a young earth.

## Shrinking Sun

Is the sun getting smaller? Some years ago an article in the **National Geographic** magazine (November, 1965) stated that the sun's mass is being reduced at the rate of four million tons per second. So, is the sun shrinking, or is it being refuelled from some source?

Some while ago, John A. Eddy (Harvard-Smithsonian Center for Astrophysics and High Altitude Observatory in Boulder) and Abram A. Boomazian (a mathematician with a Boston firm), after examining data gathered over four centuries, concluded that the sun has been shrinking at the rate of about 0.1% per century, which is equivalent to about five feet per hour.[52] Incidentally, Eddy was not a creationist.

If the rate of the sun's contraction has been constant, and one extrapolates backwards, it is reasonable to conclude that the sun must have been *larger* in mass many eons ago (if such there were).

The problem with this view is that there is considerable controversy as to whether or not the sun is, in fact, shrinking. Present data are very difficult to interpret in this regard. It does, however, remain a viable possibility if the sun's energy is the result of a "gravitational collapse." This idea is almost totally rejected by modern astronomers due to the fact that a vast age is postulated for the solar system, and this concept "shakes up" that view. It must be conceded that astronomical views are far from settled.

However, if Dr. Eddy's calculations and conclusions were accurate, at the point of 100,000 years ago, the sun would have been double its present size. Remember, according to evolutionary calculations, man had already been on earth for more than 3 million years at this time. Furthermore, at twenty million years B. C., the sun would have been touching the earth. Yet, according to the evolutionary time scale, dinosaurs already had been roaming the earth for more than 40 million years at that time!

The rate of the sun's possible shrinkage is negligible if one postulates an earth-sun relationship of only a few thousand years. The problem is insurmountable if one speculates that the earth and sun are billions of years old.

Other lines of evidence relative to the composition of the sun have also led some scientists to conclude that this "star," that so blesses our earth, is relatively "young." It does not reveal the characteristics of a multi-billion-year-old star; rather growing evidence appears to indicate that the sun "shows the characteristics of an exceedingly young star." This issue was discussed in the July, 1996 issue of **Impact**, a journal published by the Institute for Creation Research.

In the final analysis, one must remember that the Second Law of Thermodynamics would lend its confirmatory testimony to the fact that the sun is burning up, or, in the words of the sacred writer, the "heavens" are waxing old – just as the earth is (Hebrews 1:10-12).

## Pleochroic Halos

Dr. Robert Gentry is doubtless the world's leading authority on radiohalos. Radiohalos are colored spheres which form when the radiation from tiny inclusions of radioactive materials, and their daughter products, damage the crystal structure around them. Gentry has analyzed more than 100,000 of these halos in granitic rocks taken from considerable depths in various parts of the earth. Dr. Gentry contends that his studies reveal that the "basement rocks" of the earth were formed rapidly and thus evidence a young earth.

Here is the kind of process of which we are speaking. If a small amount of Plutonium (P-210, P-214, or P-218) is trapped in a molten material, and if the crystals form rapidly, the plutonium will give a little "burn out" in the rock, called a "pleochroic halo." If the rock had crystallized slowly over thousands of years, no halo would have been formed. Since they are present, they indicate a rapid creation process. Gentry affirms that the presence of these halos gives "startling evidence of a young earth."[53]

## Evidence from the Moon

It is conceded (by evolutionists and creationists

alike) that the earth and the moon are approximately the same age. So, if it can be demonstrated that the moon evidences a "youthfulness," then it will necessarily follow that the earth is young also. Thus, according to the admissions previously introduced, evolution could not have occurred.

The Bible, of course, teaches that the earth was created on the first day of the creation week and the moon was made three days later (Genesis 1:1, 14-19). On the other hand, naturalistic theories concerning the method and time of the moon's formation have been both unusual and diverse.

Since the Apollo moon explorations (1969-72), the scientific community has been totally perplexed regarding such matters. The following confession is typical of the literature available:

> "Post-Apollo studies have shown that there is no overall explanation for the origin of the planets and their moons. Some scientists feel that they may never completely learn the origin and history of the earth's immediate neighbor."[54]

Despite such acknowledgements, the popular "scientific" view is that the moon and the earth were formed at about the same time, more than 4.5 billion years ago.[55] Consider, however, the following facts concerning the moon.

Each year meteors of various sizes crash into the atmosphere of both the earth and the moon, and settle to the surface in the form of meteoritic dust. It has been estimated that anywhere from 14 million to 50 million tons of this dust descends annully.[56]

Now here is a crucial problem. If such has been occurring for nearly *five billion years*, there should be a very thick layer of meteoritic dust upon the surface of these bodies. Of course, no such layer of dust exists.

Prior to the Apollo explorations, however, it was contended that the absence of meteoritic dust upon the earth was due to erosion factors. It was speculated, though, that upon the moon, where such harsh conditions do not exist, the dust would be quite deep. In 1959, atheistic writer Isaac Asimov predicted that meteoritic dust upon the moon was at least *fifty feet* deep.[57] In fact, before the first moon landing, some scientists feared that the space craft would sink into the dust and never be heard from again. Now, however, as every schoolboy knows, the moon had only a few centimeters of dust (remember the moon tracks?). Thus, meteoritic dust obviously had not been falling upon the moon for billions of years. The evidence is that the moon is young.

Another argument for a young moon, popularized by Dr. Paul Ackerman, in his book, **It's a Young World After All** (Baker, 1986), has to do with the sharp and rugged nature of the craters on the moon (which evolutionists believe were formed early in lunar history). Ackerman has pointed out that both liquids and *solids* have something in common – they both possess viscosity (i.e., they flow downward due to gravitational attraction). Water flows quickly, honey flows less quickly, and yes, rocks flow too, though much too slowly to be detected by the human eye. The truth of this can be demonstrated. The window panes in a century-old house will be measurably thicker at the bottom than at the top.

Some rocks are harder, hence flow slower than others. The rock samples brought back from the lunar explorations are of a "soft" type (similar to basalt upon the earth). These should have eroded (flowed downward) to a far smoother state than is characteristic of them, if *the moon were several billion years old.* But they haven't.

Studies by Harold Slusher and others at the University of Texas at El Paso have suggested that the "lunar craters cannot be more than a few thousand years old."[58] There is something obviously wrong with the way the moon has been dated; it is not as old as evolutionists surmise.

There are numerous other evidences from the material world that argue that the earth is *not* billions of years old. The creationist would be well advised to master a few of these arguments in his efforts to teach those who have faith problems due to the influence of evolution.

# 10 Arguments for an Ancient Earth Answered

Evolutionists stress that there are a number of earth-features that have required vast amounts of time, and therefore, on the basis of these phenomena, our globe must be billions of years old. We will address a few of these popular items.

## Dating Techniques

It is argued that the radiometric dating processes have demonstrated that the earth is ancient. We only allude to this argument at this point due to the fact that it should be considered in a chapter of this nature. However, we have already dealt with the assertions of evolutionary chronologists in Chapter 3. There, we observed that all of the radiometric dating techniques are based upon certain uniformitarian (evolutionary) assumptions. In other words, the clock systems were designed with evolution in mind. (If you allow a person to design his own clock, he can run the mile in *ten seconds!)* This is evidenced by the fact that the alleged great age of the earth was *already fixed* by evolutionists long before the dating methods were even invented. Let me emphasize again, there is absolutely no hardcore, scientific evidence (radiometric or otherwise) that demands an earth that is billions of years old.

## The Evolutionary Process

It is contended that the evolutionary process is very slow, and, since it is evident that evolution has occurred, the earth *must be* quite old. What a marvelous argument – a complete stranger to logic! This is a classic example of circular reasoning. The earth must be old because evolution is true, and evolution has occurred because an ancient earth provides the time for such. Such an argument hardly needs the dignity of a response.

## Stalactites and Stalagmites

Sometimes water, with mineral deposits, will drip steadily in caves, etc., and slowly form limestone "icicles" called stalactites (if they hang) or stalagmites (if they form upward). The Carlsbad Caverns in New Mexico, and Mammoth Cave in Kentucky, are examples of caves with these formations.

It is generally assumed that it takes about one hundred years for the formation of one inch of a stalactite or a stalagmite. Accordingly, it is alleged, millions of years must have been required for such huge formations to build. Carlsbad Caverns are said to have begun forming about sixty million years ago. Again, though, there is much evidence that indicates that, under proper conditions, stalactites, etc., can build rather rapidly.

Some years ago, the **National Geographic** magazine published a photograph of a bat that had fallen onto a stalagmite in Carlsbad Caverns. The stalagmite had grown so fast that it preserved the bat before it had time to decompose.[59]

A "curtain" of stalactites has formed from the foundation of the ceiling beneath the Lincoln Memorial in Washington, D. C. (constructed in 1923). Some of these are five feet in length. At the uniformitarian rate suggested above, that would indicate that the Lincoln

Memorial was built about *six thousand years ago*.

In an unused wing of the Milwaukee Public Museum, which is underground, there are stalactites some six feet long. This would suggest that the Museum was built some seventy-two hundred years ago, or some *five thousand, seven hundred years before Columbus sailed to the new world.*[60]

## Petrification

Petrification is the process of the changing of plant and animal bodies (or parts) in the earth into stone, or hard, stone-like material, as the result of the operation of water-containing chemical substances gradually replacing the tissue. Frequently, though, the objects are preserved in the very same structure format. It is generally supposed that the process takes millions of years to occur.

Mr. H.G. Labudda of Kingaroy in south-east Queensland (Australia) specializes in collecting petrified objects. Among the articles of his collection is a perfectly petrified orange. It was made in a creek at nearby Gayndah, an area known for its oranges.

However, Gayndah was not founded until 1849, and oranges were not raised in the area until 1868.[61]

This writer has seen and photographed a petrified apple near Oakdale, California. The owner of this apple told the following story of its discovery. A young boy planted several apple trees in the 1950's as a high school agricultural project. Less than twenty years later, the field was disked around the trees and five petrified apples were discovered. The author's photograph of one of these apples appeared in the August 1975 issue of the **Bible-Science Newsletter**. Here was an object that had petrified within a very short period of time. Petrification does not require millions of years.

## Formation of Coal and Oil

There are various kinds of coal, ranging from the hard anthracite to the soft lignite. Coal is formed from various materials – primarily, however, from wood tissue. Evolutionists allege that the great coal beds of the earth were laid down as a consequence of the gradual (millions of years) accumulation of vegetable matter in bogs and marshes. Supposedly, it takes about one thousand years to produce one inch of coal. (Note: It is interesting to reflect upon the fact that there are many kinds of trees and plants that *do not grow in swamps and bogs* that are found in coal deposits.)

Actually the evidence does not support the evolutionary view. The geological data strongly indicate that the vegetation involved in the formation of coal beds was collected by great water currents dumped from various locations and covered with sediment. This produced intense pressure and heat which caused the vegetable matter to convert to coal in a rather rapid fashion.

In many places there are fossilized trees spanning several coal layers. This indicates that the surrounding coal was formed so quickly that termites did not have time to consume the wood! E. S. Moore, a coal geologist, declares: "From all available evidence it would appear that coal may form in a very short time, geologically speaking, if conditions are favorable."[62]

> "Near Freiburg, Germany, a certain wooden railroad bridge was being replaced with steel. The wooden piles which had supported the weight were found to have turned partially into coal. This proves that

coal can form within the short hundred year span of the railroad era ... A coal-like substance similar to anthracite has actually been produced in the laboratory by enclosing vegetation in a steel bomb and subjecting it to heat and pressure."[63]

As a matter of fact, at Spirit Lake near Mount St. Helens in Washington, an enormous log mat which floats on the Lake (as a result of the May, 18$^{th}$, 1980 eruption) has lost its bark and branches as a result of wind and waves. Much of the debris has sunk to the bottom of the lake and has formed a layer of peat several inches thick. This peat resembles, both compositionally and texturally, certain coal beds of the eastern United States. It is believed that this may be the first stages of the formation of coal at this site.[64] Coal forming in only eight years!

Petroleum is the result of a complex mixture of organic compounds which have been buried in sediments and undergone chemical change. The same uniformitarian assumptions are made with reference to petroleum as with coal. But the United States Bureau of Mines at Pittsburg has produced oil experimentally in only one hour! Refuse was subjected to two hundred and fifty degrees centigrade and only fifteen hundred pounds of pressure, processed about an hour, and allowed to cool. One hundred pounds of garbage can produce about two gallons of oil.[65]

A very strong case can be made for the production of most oil and coal in conjunction with the Genesis Flood.

## Starlight from Distant Bodies

Light travels at the rate of 186,000 miles per second, or about 6 trillion miles in one year. At this rate, it is believed that it would require some 5 billion years for light (which we see) to reach the earth from some of the distant galaxies. It thus is claimed the universe must be quite old. It should be recognized that there is considerable speculation in astronomical calculations, and much about outer space is yet unknown. In response to the above, these points are suggested for further study.

First, there are biblical indications that the universe was created fully functional, which demands a certain level of maturity. Trees were growing, rivers were flowing, Adam and Eve were adults, etc. Stars were created to functionally provide lights for the earth (signs and seasons, etc. – Genesis 1:14) and to be a testimony to the glory of God (Psalm 19:1). Obviously, the light from stars was visible upon the earth from the beginning. At the time of Abraham, only a few thousand years from the original creation, the stars, as viewed from the earth, were innumerable (Genesis 15:5). Some feel, however, that this does not fully explain the astronomical picture, hence, other possibilities are suggested as well.

Some studies have appeared to indicate that light may take a "shortcut" as it travels in deep outer space. If this theory has validity, it would suggest that it does not take light nearly as long to reach the earth from outer space as is believed commonly. By the use of certain mathematical calculations, for example, it is speculated that light could reach the earth from the most distant known star in only 15.72 years.[66]

A rather new theory that has attracted some attention is that the speed of light has not always been constant. Australian scientist Barry Setterfield has argued, based upon data collected over the past three centuries, and mathematically extrapolated backwards, that the speed of light was five billion times faster just six thousand years ago. If such were true, light from the farthermost star could have reached the earth in only three days.[67] Though some have disputed Setterfield's theory, according to an article which appeared in Britain's prestigious journal, **Nature** (August 8, 2002), some evolutionists now are claiming that "light has been slowing down since the creation of the universe."[68]

It is virtually certain that many of the mysteries of outer space will never be understood by man; certainly it is likely that purely naturalistic explanations will always be highly speculative and subject to constant revision. This is the track-record of "science." Today's "science" is tomorrow's superstition.

The Bible contains two types of information – that which is "checkable" and that which is not. When the checkable information of Scripture consistently proves to be true, one should not hesitate to accept its testimony in areas where it merely describes a circumstance that is beyond man's current ability to document. Scientific speculation is not a valid reason for rejecting the statements of the Bible. When all of the evidence is not in, it is better to trust God's Word – and wait.

## The Grand Canyon

The Colorado River is approximately fourteen hundred miles long. It begins in Wyoming and ultimately spills into the Pacific Ocean at the Gulf of California (Mexico). As it meanders toward its goal, it passes through one of the most spectacular places on earth, the Grand Canyon. This yawning chasm is four hundred and seventeen miles long, and it is anywhere from fourteen to eighteen miles wide. It has a maximum depth of five thousand, seven hundred feet.

It is believed, of course, by those who subscribe to uniformitarian geology (i.e., that natural degenerative processes have occurred quite slowly and at relatively constant rates of decomposition) that the Colorado River carved the Grand Canyon in the western wilderness over a vast era of history. To the average person, it does appear that this huge rift in the earth's surface must have required much time.

Professor Ronald Ives of Northern Arizona University has written:

> "Formation of the Grand Canyon took place in two major steps: the deposition of the beds that were later eroded into a gorge – a process lasting considerably more than one billion years – and erosion of the canyon proper, which required not more than 10 million years."[69]

What evolutionists refuse to consider, however, is that a catastrophe, such as a great flood, could have constructed the features of the canyon in a relatively short period of time. Among the evidences supporting such a hypothesis is the data being gathered since the eruption of Mount St. Helens.

On May 18,1980, Mount St. Helens, a volcano in Washington state, exploded. There was a release of steam equivalent to twenty million tons of TNT. One hundred and fifty miles of forest was wiped out in six minutes. Huge rock slides, water waves (at nearby Spirit Lake), steam output, etc., generated an energy release roughly equivalent to twenty thousand atomic bombs like those employed at Hiroshima during World War II.

The effects of this "local" catastrophe have been carefully studied over the past twenty-plus years, and some surprising information has resulted.

First, in 1986 it was reported that new strata, over six hundred feet thick, have formed at Mount St. Helens since 1980.[70] That is an average of one hundred feet per year. One deposit accumulated twenty-five feet in less than one day! If one postulates that a similar (or greater) catastrophe was instrumental in forming (partially or completely) the Grand Canyon, and that a comparable rate of deposition was involved, at the rate of one hundred feet per year, the five thousand, seven hundred feet strata of the Grand Canyon could have been laid down in fifty-seven years, and thus not have required that "considerably more than a billion years" speculation.

Second, there have been numerous subsequent mudflows as an aftermath of the St. Helens' eruption. For example, on March 19, 1982, a canyon, which has come to be known as "the little Grand Canyon," was formed in one day – one hundred and forty feet deep! The "little Grand Canyon" is a one fortieth scale model of the "big Grand Canyon." Thus, conceivably, the Arizona Grand Canyon could have been catastrophically fashioned in only *forty days!* Evidence suggests, therefore, that the Canyon could have been formed much quicker than the ten million years argued by evolutionists.

## Coral Reefs

The coral is a marine organism that lives in colonies, usually where the water is warm and shallow. Each coral builds a skeleton of limestone around it, taking the limestone from seawater. When the coral dies, it leaves its skeleton behind; these skeletons can build upon one another, eventually forming rocky ridges called a coral reef.

The rate of growth in coral is not yet known, but it is generally assumed that coral reefs grow very slowly. Some reefs near Key West, Florida are estimated to grow at the rate of about one half-inch per year. Large reefs, therefore, presumably argue for the great age of the earth. However, two factors need to be considered.

First, there is some evidence that coral structures may grow at a much faster rate than is normally assumed – if the environment is favorable. For example, underwater explorers recently found a five-foot diameter coral growth on a bow gun of a ship sunk in 1944.[71]

Second, there is considerable evidence which suggests that these reefs are not composed solely of coral growth; rather, there is much *inorganic* material combined within the build-up. Thus Morris points out that

> "the evidence indicates that coral and other fossil organisms were simply transported into place by sedimentary processes, with the lime muds which constituted the matrix of transport later becoming lithified to form the present fossiliferous limestones."[70]

Although there is much study yet to be done on this subject, there is no reason for assuming that coral reefs, either living or fossil, require more than a few thousand years for their production.

# 11 A Concluding Word: Where is the Evidence?

Let us take a moment to summarize what we have learned from the material of this book.

First, we have observed how very crucial "time" is to the evolutionary plot. It is *everything!* Evolutionists readily admit that without it they are utterly and hopelessly lost. But we also noted that time has no power; it is simply there. The truth is, time, accommodates degeneration. It is not the hero of the story.

Second, it has been stressed that there is absolutely no scientific evidence that *proves* the earth is billions of years old. The dating techniques that are employed to yield the high numbers for the earth's age are all grounded upon uniformitarian assumptions that take for granted that evolution is true. There are many strong reasons why the assumptions of the dating methods should be questioned.

Third, we introduced several examples of dates yielded by the radiometric systems which are in great conflict with one another, and which, reveal, therefore, that there is something drastically wrong with these methods. The evolutionary "clocks" are sadly in need of repair.

Fourth, we showed that the subject of "chronology" is not something the Bible ignores; it is not an irrelevant issue. God has filled much of his Book with chronology. If there are chronological considerations that shed some light upon the antiquity of the earth and man, we need to respectfully accept the testimony of the Scriptures, and wait for science to "catch up" with inspired revelation.

Fifth, we established that the genealogies of the biblical record do provide a general chronological framework with which to determine the relative infancy of the human family. While there may be some level of elasticity in some of the genealogical records, they must of necessity be relatively minor. Biblical chronology argues against the notion that man has been upon the earth for millions of years.

Sixth, it was made very clear, by the use of several passages, that the earth and human-kind were created in the same week. It is thus apparent that the earth can not be billions of years old. Only by ignoring the plainest of biblical teaching can one argue that man is a recent "newcomer" to this planet.

Seventh, we noted that, unfortunately, many Christians have been woefully intimidated by evolutionary chronology, and that they, therefore, yielding to the pressures of this philosophy, have concocted bizarre theories. These include the Gap Theory, Day-Age Theory, etc., in an attempt to bring about a harmony between the Scriptures and the theory of evolution. But such theories are without biblical support and they are rejected by zealous evolutionists and sound Bible scholars alike.

Eighth, it was observed that if the scriptural record of earth and human history is true, one should expect to find certain physical evidences that support this affirmation. And we discovered that there are many evidences indeed which buttress the biblical claim that this planet and its inhabitants are only several thousand years old, not millions or billions of years.

Ninth, we looked at some of the popular arguments that are employed to support the notion that the earth is very ancient. We showed, on the basis of physical evidence and logical reasoning, that these arguments are far from conclusive. They do not make the case for a billions-of-years-old earth.

Accordingly, all of this evidence, taken together, makes a tremendously powerful case against the theory of evolution. It reveals that much of what is paraded today under the banner of "science" is quite the opposite. It is mere superstition and anti-religious bias.

We can think of no better way to conclude this discussion than to introduce a couple of rather well-known quotations, which somehow seem appropriate.

Mark Twain was an irreverent, old gentleman – certainly no friend of the Bible. Yet he had enough insight into some of the scientific nonsense of his day to poke fun at the evolutionary speculations with which he was contemporary. In his **Life on the Mississippi**, he wrote:

> "In the space of one hundred and seventy-six years the Lower Mississippi has shortened itself two hundred

and forty-two miles. That is an average of a trifle over one mile and a third per year. Therefore, any calm person, who is not blind or idiotic, can see that in the Old Oolitic Silurian Period, just a million years ago next November, the Lower Mississippi River was upward of one million three hundred thousand miles long, and stuck out over the Gulf of Mexico like a fishing-rod. And by the same token any person can see that seven hundred and forty-two years from now the Lower Mississippi will be only a mile and three quarters long, and Cairo and New Orleans will have joined their streets together, and be plodding comfortably along under a single mayor and a mutual board of aldermen. There is something fascinating about science. One gets such wholesale returns of conjecture out of such trifling investment of fact."

Then consider this amusing quote from Lewis Carroll's **Alice In Wonderland**.

"Alice laughed. 'There's no use trying,' she said, 'One can't believe impossible things.' 'I dare say you haven't had much practice,' said the Queen. 'When I was your age I did it for half an hour a day. Why sometimes I've believed as many as six impossible things before breakfast.'"

One wonders if the Queen had in mind the theory of evolution!

# Endnotes

1. See the author's article, "The Evolution Revolution," *Essays in Apologetics*, III (Montgomery, AL: Apologetics Press, Inc., 1988), pp. 89-96.
2. Robert Jastrow, *Until the Sun Dies* (New York: Warner Books, 1980), p. 112.
3. George Wald, "The Origin of Life," *Scientific American* (August 1954), pp. 45-53.
4. George G. Simpson, *The Major Features of Evolution* (New York: Columbia University Press, 1953), p. 96.
5. Stephen Moorebath, *Scientific American* (March 1977), p. 92.
6. John Eddy, *Geotimes* (September 1978), p. 18.
7. Robert Kofahl, *Handy Dandy Evolution Refuter* (San Diego, CA: Beta Books, 1980), p. 109.
8. Donald Chittick, *A Symposium on Creation II* (Grand Rapids, MI: Baker, 1970, p. 73).
9. Henry M. Morris, *The Biblical Basis of Modern Science* (Grand Rapids: Baker, 1984), p. 267.
10. David E. Seidemann, *Geological Society of America Bulletin* 88 (November 1977), 1660.
11. Morris, *op. cit.*, p. 266.
12. Frederic B. Juenemann, *Industrial Research and Development* (June 1982), p. 21.
13. William S. Beck, *Modern Science and the Nature of Life* (New York: Harcourt, Brace, 1957), p. 170.
14. Kenneth L. Currie in: *Rock Strata and the Biblical Record*, Paul Zimmerman, Ed. (St Louis: Concordia, 1970), p. 70.
15. E.H. McKee & D.C. Noble, "Age of the Cardenas Lavas, Grand Canyon Arizona," *Geological Society of America Bulletin* 87 (August 1976), pp. 1188-1190.
16. *Nature Physical Science*, Vol. 232 (July 19, 1971), pp. 60-61.

17  A. Hayatsu, *Canadian Journal of Earth Sciences* 16 (1979), 974.

18  Wayne Jackson, *The Mythology of Modern Geology* (Stockton, CA: *Apologetics Press*, Inc., 1980).

19  Funkhouser & Naughton, *Journal of Geophysical Research (July* 15, 1968), p. 4601.

20  Harold W. Clark, *The Battle Over Genesis* (Washington, D. C.: Herald & Review, 1977), p. 138.

21  W. F. Libby, *American Scientist,* Vol. 44 (January 1956), p. 107.

22  Froelich Rainey, "Dating the Past," 1971 *Yearbook of Science and the Future* (Britannica), pp. 390-391.

23  M.S. Keith & G.M. Anderson, *Science* (August 16, 1963), p. 634.

24  Dort, *Anartic Journal of the U. S.,* 6 (1971), p. 210.

25  Robert Brown, *Review & Herald,* Vol. 148, No. 44 (October 28, 1971).

26  Caryl Haskins, *American Scientist,* Vol. 59 (May/June 1971), p. 298.

27  C.W. Ferguson, *Science* 159 (February 23, 1968), p. 840.

28  Donald England, *Faith and Evidence* (Delight, AR: Gospel Light, 1983), p. 155.

29  Edwin Theile, *A Chronology of the Hebrew Kings* (Grand Rapids: Zondervan, 1977), p. 7.

30  John N. Clayton, *Does God Exist? Course* (Lesson 4), p. 3.

31  Clayton, *op. cit.,* (Lesson 8), p. 2.

32  J.D. Douglas, Ed., *The New Bible Dictionary* (Grand Rapids: Eerdmans, 1974), p. 213.

33  J. Barton Payne, *Zondervan Pictorial Bible Encyclopedia* (Grand Rapids: Zondervan, 1975), Vol. 1, p. 831.

34  John W. Klotz, *Genes, Genesis and Evolution* (St. Louis: Concordia, 1970), p. 91.

35  H. Cremer, *Biblico-Theological Lexicon of New Testament Greek* (Edinburgh: T. & T. Clark, 1962), pp. 113-114, 381.

36 S.T. Bloomfield, *The Greek New Testament With English Notes* (Boston: Perkins & Marvin, 1837), 1, pp. 197-198.

37 Arndt & Gingrich, *A Greek-English Lexicon of the New Testament* (Chicago: University of Chicago Press, 1967), p. 86.

38 *Ibid.*, p. 446.

39 W.E. Vine, *Expository Dictionary of New Testament Words* (Westwood, NJ: Fleming Revell, 1962), IV, p. 233.

40 R.C. Trench, *Synonyms of the New Testament* (London: Kegan, Purl, Trench, Trubner & Co., 1890), pp. 215-216.

41 J.H. Thayer, *Greek-English Lexicon of the New Testment* (Edinburgh: T. & T. Clark, 1958), p. 452.

42 James McKnight, *Apostolical Epistles* (Nashville: Gospel Advocate, 1960), p. 58.

43 Oswald T. Allis, *God Spake By Moses* (Nutley, NJ: Presbyterian & Reformed, 1958), p. 10; cf. also 153ff.

44 Raymond Surburg in: *Darwin, Evolution, and Creation*, Paul Zimmerman, Ed. (St. Louis: Concordia, 1961), pp. 53-54.

45 Harold Stigers, *A Commentary on Genesis* (Grand Rapids: Zondervan, 1976), p. 49.

46 Alexander Campbell, *Lectures on the Pentateuch* (Rosemead, CA: n.d.), p. 96.

47 Henry Morris, *The Biblical Basis for Modern Science*, p. 416ff. (Cf. Note 7.)

48 Fritz Heide, *Meteorites* (Chicago: University of Chicago Press, 1964), p. 119ff.

49 Thomas G. Barnes, "Depletion of the Earth's Magnetic Field," *ICR Impact* (October 1981).

50 Paul Steidl, *The Earth, the Stars, and the Bible* (Nutley, NJ: Presbyterian & Reformed, 1979), p. 30.

51 Melvin Cook in: *Bible-Science Newsletter* (January 15, 1970).

52 Russell Akridge, "The Sun is Shrinking," *ICR Impact* (April 1980).

53 Robert Gentry, *Annual Review of Nuclear Science, Vol. 23* (1973), p. 247.

54 "The New Moon," *Time* (April 8, 1974), p. 52.

55 J.A. Wood, "The Moon," *Scientific American* (September 1975), pp. 92-102.
56 Hans Pettersson, *Scientific American* (February 1960), p. 132, G. S. Hawkins in: *An Introduction to Planetary Physics* (New York: Wiley & Sons, 1968), p. 249.
57 Isaac Asimov, *Science Digest*, (January 1959), pp. 33-36.
58 Glen Morton, Harold Slusher, Richard Mandock, "The Age of Lunar Craters," *Creation Research Society Quarterly* (September 1983), pp. 106-107.
59 *National Geographic* (October 1953), p. 442.
60 *Bible-Science Newsletter* (June 1983), p. 10.
61 John Osgood, "Rock Hard Orange," *Ex Nihilo*, Vol. 10, No. 1 (December 1987 – February 1988), p. 11.
62 E. S. Moore, *Coal* (New York: Wiley, 1940), p. 143.
63 Reginald Daly, *Earth's Most Challenging Mysteries* (Nutley, NJ: Craig Press, 1972), pp.138-139.
64 Steven Austin, "Mount St. Helens And Catastrophe," *ICR Impact* (July 1986).
65 Daly, *op. cit.*
66 Richard Niessen, "Starlight And the Age of the Universe," *ICR Impact* (July 1983).
67 Barry Setterfield, "The Velocity of Light and the Age of the Universe," *Ex Nihilo* 1 (1982), pp. 52-93.
68 P.C.W. Davies, T.M. Davis, & C.H. Lineweaver, "Black Holes Constrain Varying Constants," *Nature* (August 8, 2002), pp. 602-603.
69 *Encyclopedia Americana,* 13 (Grolin Inc., 1985), pp. 163-164.
70 Austin, *op. cit.*
71 Sylvia Earl, *National Geographic* 149 (May 1976), pp. 578-613.
72 Henry Morris, *Science, Scripture and the Young Earth* (El Cajon, CA: Institute for Creation Research, 1983), p. 9.

# Appendix

## Why People Believe in Evolution

The most insidious and damaging ideology ever foisted upon the mind of modern man is the notion that human beings are but animals, and the offspring of other, more primitive creatures. It is known as the theory of organic evolution. This concept has been reflected in recent years in such volumes as Phil Donahue's, *The Human Animal* (1986), and in the earlier production, *The Naked Ape* (1967), (as man was characterized) by zoologist, Desmond Morris.

Tragically, multiplied thousands have ingested, to a greater or lesser degree (sometimes even with a religious flavor), this nefarious dogma. But why? Have folks intellectually analyzed the matter, and thus, on the basis of solid evidence and argument, accepted this viewpoint. Not at all; rather, for a variety of emotional reasons, this concept is entertained so readily.

In 1974, Marshall and Sandra Hall published a

book titled, *The Truth: God or Evolution?* In the opening section of this excellent volume, the authors listed several reasons why the evolutionary theory is embraced by so many. With credit to them for the germ thoughts, I would like to expand the discussion.

## Brainwashing

Since the issuance of Charles Darwin's, *The Origin of Species* (1859), there has been a massive campaign to flood the "intellectual market" with evolutionary propaganda. Though such ideas by no means originated with Darwin, he popularized evolution more than anyone else. His book sold out (1,025 copies) the first day of its release.

Another significant milestone was the famous Scopes Trial, conducted in Dayton, Tennessee in July of 1925. Twenty-four year old John Thomas Scopes, a high school science teacher, had agreed to violate Tennessee's Butler law, which forbade the teaching of any theory that holds man has descended from a lower form of life. The entire affair was "rigged," but it brought together William Jennings Bryan (three-time

Democratic nominee for president), who volunteered to represent the state, and the famed criminal defense attorney, Clarence Darrow, who defended Scopes. The trial, the first ever to be broadcast on radio, brought national attention to the issue of creation vs. evolution. As a result of that encounter, creationism was cast in an unfavorable light, and evolutionary dogma gained considerable respectability, albeit undeserved.

From that time, however, the theory of evolution has accelerated in influence via the media and the public school system. Today, there exists a determined campaign for the indoctrination of evolution, and millions have absorbed it into their minds.

## Intimidation

Hand-in-hand with the brainwashing factor is the impact of intimidation. Supposedly, evolutionary doctrine has the endorsement of "science." In 1966, H. J. Muller, a prominent geneticist, circulated a statement signed by 177 biologists. It asserted that evolution is a "scientific law" which is as firmly established as the rotundity of the earth.

Since most folks want to be thought of as "educated," and as they have been led to believe that "all educated people believe in evolution," they have defected to the Darwinian camp. Most of these individuals could not cite a solitary argument in defense of evolution; they simply believe it is fact because "the scientists say so."

Informed people should know the following:
1. Evolution is not a scientific law. It is a mere hypothesis that falls quite beyond the pale of the scientific method (observation, experimentation, and verification).
2. There are numerous laws, e.g., the laws of thermodynamics, genetics, etc., which contradict evolutionary assertions.
3. Many scientists dispute that evolutionary dogma is true science. Evolutionist Robert Jastow, for example, has conceded that belief in the accidental origin of life is "an act of faith," much, he says, like faith in the power of a Supreme Being (*Until the Sun Dies*, New York: Warner Books, 1977, p. 52).

Theodore N. Tahmisian, a nuclear physicist with the Atomic Energy Commission, has said:

> "Scientists who go about teaching that evolution is a fact of life are great con men, and the story they are telling may be the greatest hoax ever. In explaining evolution we do not have one iota of fact ... It is a tangled mishmash of guessing games and figure jaggling ... If evolution occurred at all, it was probably in a very different manner than the way it is now taught" (*Fresno Bee*, Aug. 20, 1959).

It is hardly necessary, therefore, to yield to the pressures of evolutionary brow-beating. We ought not to be cowed down; we should be more aggressive, demanding that those who affirm their confidence in evolution argue their case logically.

## Religious Confusion

Some have been thrust toward evolutionary ideology because they are repelled by the confused (and sometimes cruel) state of the religious world. Religionists have sacrificed their own children in the name of "gods" (cf. Jer. 19:5). In the Far East, the cobra is worshipped as deity. "Christians" (so-called) have warred with the devotees of Islam.

Catholics allege that the bread and wine of "the Eucharist" magically turn into the body and blood of Jesus, while Protestants insist that such does not occur. Some contend that "baptism" is administered only by immersion, while others allege that "sprinkling" or "aspersion" will suffice. A rather unique view suggests that it takes all three "modes" to constitute the "one baptism" of Ephesians 4:5 (cf. *Wycliffe Bible Dictionary*, Peabody: MA: Hendrickson, 1998, p. 201).

This disunity has driven many to disenchantment with religion in general, which includes a rebellion against divine revelation. This, of course, is precisely what Jesus indicated. He admonished those who professed a loyalty to him to be "one," that "the world might believe" (Jn. 17:20-21); the Lord thus implied that disunity would produce the opposite effect, i.e., unbelief.

People need to realize that a departure from the original does not negate its genuineness. The segmented status of "religiondom" does not authenticate evolution. The fact of the matter is, the evolutionists are as divided as the religionists.

For example, Sir Francis Crick, co-discoverer of DNA, contended that biological life evolved here on earth. On the other hand, Sir Fred Hoyle has argued that "spontaneous generation" occurred in outer space! Some Darwinians speculate that the evolutionary process has occurred quite gradually, over eons of time. Supposedly this explains the lack of transitional forms in the fossil record. Others (e.g., Richard Goldschmidt, and more recently, Stephen Gould of Harvard), suggest that evolution has proceeded rapidly, almost in snatches.

There is wholesale disagreement among the advocates of evolution. Those, therefore, who have fled from religion because of its disunity, have found no haven in Darwinism.

## A World of Disorder

Many feel that our world environment, which is so characterized by brutality and suffering, is more consistent with Darwin's tooth-and-claw, "survival-of-the-fittest," principle, than it is with the notion that the earth is tended by a benevolent God. There might be some leverage in this argument if there were no other rational explanation for the ills of this globe.

But the fact is, a compelling case can be made for the proposition that life's tragedies are the result of man's rebellion again his Creator; and negative consequences have been allowed to follow as an educational process on behalf of the human family. In our recently published book, *The Bible and Mental Health*, we have an entire chapter chronicling some of the values of human affliction.

But here is another matter for consideration. While the believer has some basis for explaining the presence of "evil" in a fashion that is consistent with the existence of a powerful and benevolent God, the evolutionist has no reasonable explanation as to why there is a human sensitivity within man that judges some things to be "evil" and others "good." How can a package of mere "matter," which, according to atheism, is the sum of man, arrive at a rational, moral judgment concerning this phenomenon called "evil"? The problem of "evil" is more challenging for the evolutionist than for the creationist.

## Tangible Evidence

Many folks are impressed with the evolutionary case because it is buttressed, they believe, with tangible evidence, whereas religion seems to partake of a dreamy, surreal environment. After all, scientists have "fossils" to prove their case, don't they?

This argument is exceptionally deceptive for the following reasons:

1. All of the fossils ever collected represent less than 1% of the potential evidence, according to David Raup of Chicago's Field Museum (*Museum Bulletin*, Jan., '79, p. 50).
2. Not a single fossil has ever been discovered that clearly demonstrates a link between basic organism "kinds."
3. All fossil evidence is subject to interpretation; and even evolutionists dispute the data.
For example, when Donald Johanson and his colleges discovered the few bone fragments they dubbed "Lucy," back in 1974, they al-

leged that this little creature walked on two legs, and was on-the-way to becoming human. Numerous evolutionists, however, seriously disputed this. We discussed this matter in considerable detail in the October, 1986 issue of the printed *Christian Courier*.

But Bible believers are not without "tangible" evidence in the defense of their case. Numerous archaeological discoveries have been made which support the historicity of the Scriptures (see our book, *Biblical Studies in the Light of Archaeology*).

If, then, a general case can be made for the factual correctness of the Bible, one may reasonably conclude that its affirmations regarding the origin of humanity are correct as well.

## Escape From Responsibility

Another reason why many so readily accept evolution as the explanation for mankind, is that such allows them to "cut loose" from God, and hence to be free from moral and religious obligations. They thus can become their own "gods," and write their own rules. Richard Dawkins says that "Darwin made it possible to be an intellectually fulfilled atheist" (*The Blind Watchmaker*, New York: W.W. Norton, 1986, p. 6).

This viewpoint was vividly illustrated some years ago when Clarence Darrow spoke to the inmates of the Cook County jail in Chicago. Hear him.

> "I do not believe there is any sort of distinction between the real moral conditions of the people in and out of jail. One is just as good as the other. The people here can no more help being here than the people outside can avoid being outside. I do not believe that people are in jail because they deserve to be. They are in jail simply because they cannot avoid it on account of circumstances which are entirely beyond their control and for which they are in no way responsible" (Arthur Weinberg, *Attorney For The Damned*, New York: Simon & Schuster, 1957, pp. 3-4; *emp. WJ*).

This shocking statement reveals the motive of some evolutionists.

## Conclusion

People do not believe in evolution because they have been led there by solid evidence. They are stampeded into the Darwinian community by superficial, emotional, and personal factors. They only delude themselves when they think otherwise.

# The Bible & Science

*by Wayne Jackson*

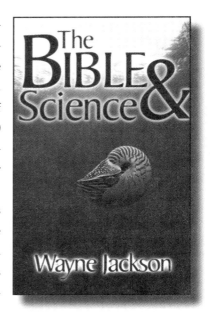

May one assert that the "spiritual" truths of the Bible are meaningful, but its "scientific" references are flawed? No, that is not consistent. The "sum" of the various parts of Sacred Writ are "truth" (Psa. 119:160 ASV). The Scriptures are scientifically credible. In fact, "science" never quite "catches up" with Scripture.

Wayne Jackson addresses whether science and the Bible agree or disagree. His style is easy to read, and yet, he brings complex issues down to a level that is easy for everyone to understand.

Parents will treasure this volume as they discuss with their children the day's schoolwork. College students assaulted by humanistic professors will find their faith in the Bible a comfort rather than something to be ashamed of. Read this book — then, give a copy to someone you love.

For more information regarding these and other materials that will strengthen your faith, visit Courier Publications at their web site or write to the address below.

**Courier Publications**
7809 N. Pershing Ave.
Stockton, CA 95207

http://www.christiancourier.com
http://www.courierpublications.com